MW01697520

Oodles of Bootles

Written by Carol Pugliano Illustrated by Larry Moore

S F
A W
Scott Foresman
Addison Wesley
http://www.sf.aw.com

CHILDREN'S TELEVISION WORKSHOP
CTW
http://www.ctw.org

ISBN: 0-201-31761-3

Printed in the United States of America.

4 5 6 7 8 9 10-BX-01 00

We are looking for bootles.
They are so sweet.
They are little bugs
that wear boots on their feet.

Each bug has 2 heads,
10 legs, and 5 tails.
They are smaller than puppies
but bigger than snails.

Bootles
5 Tails
2 Heads
10 Legs

Four tiny bootles
come out every spring.
They have a good time.
They dance and they sing.

We will give them a party.
It will be such fun!
We will give bootle gifts
to each little one.

We need 1 hat
for each bootle head.
How many in all?
"Count by 2s," Herman said.

We need 1 ribbon
for each bootle tail.
How many in all?
Count by 5s. You won't fail!

Each bootle foot
needs 1 shiny red boot.
Count by 10s for the answer.
Won't the bootles look cute?

We are ready to dance.
We are ready to sing.
Where are the bootles?
Are they ready for spring?

8 Hats
20 Ribbons
40 Boots
Bootles Party

Hip hip hooray!
The bootles are here.
Let us dance. Let us sing.
Let us give them a cheer!

We have ribbons and boots
and a hat for each head.
The bootles are happy.
"Thank you!" they all said.

We are having fun.
Now here's something new.
Here are 6 more bootles.
They will join the party too.

We need ribbons and boots
and hats for each one.
Can you help us count them
so we can have fun?

The HYRAX An Interesting Puzzle

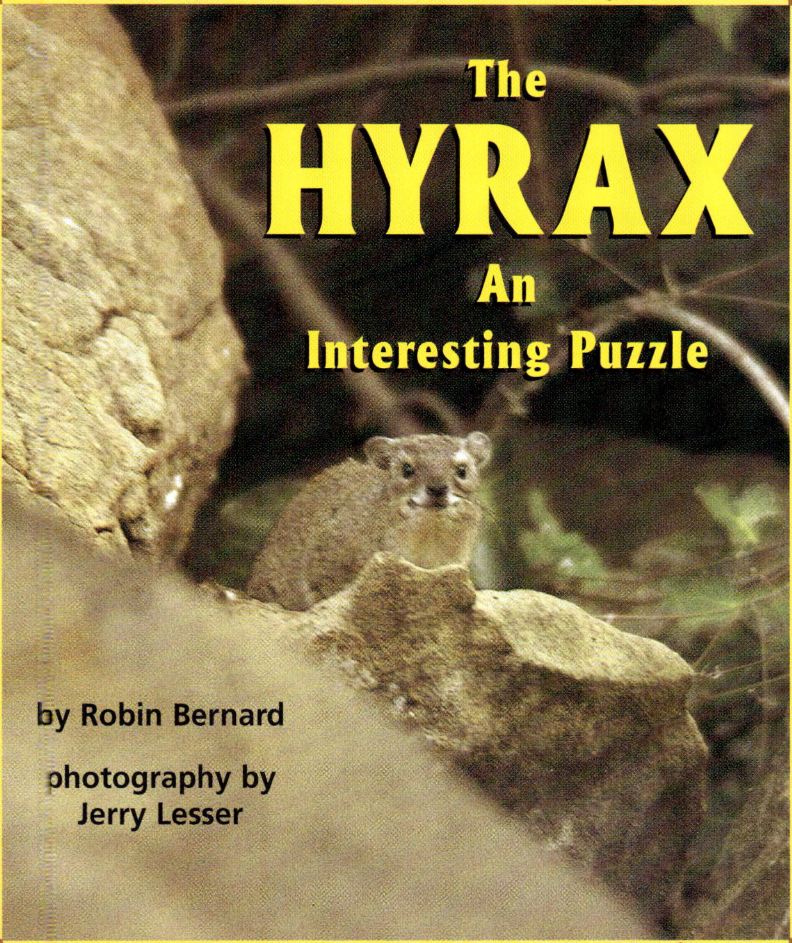

by Robin Bernard

photography by
Jerry Lesser

HOUGHTON MIFFLIN BOSTON

You can't smell the ocean from Top-Knot Island. You can't see waves or hear the gulls. Top-Knot Island is a rocky island. But it isn't like other islands, because it's in the middle of a sea of grass on the flat plains of Africa.

Top-Knot Island is really a group of boulders. It's like an island in some ways. It has its own climate, its own plants, and its own animals.

I went there to study an animal most people have never heard of — the hyrax.

I saw my first hyrax on an earlier trip to Africa. I climbed some rocks to take pictures of lizards and I had the feeling that I was being watched.

I was right! A furry little animal was looking at me. I took its picture before it ran away.

It came back a few minutes later, and it brought some friends. They all lined up to stare at the strange two-legged animal with the camera. They had funny smiles and bright eyes. I wanted to find out more about them.

Hyraxes are not well-known animals. Perhaps this is because hyraxes don't look or act special like some other animals. They don't have long, beautiful horns or lovely fur. In fact, they look very ordinary.

But scientists find hyraxes interesting. The hyrax has the same kind of stomach as a horse. It also has many different kinds of teeth. Some of its teeth are like hippo teeth. Some are like rhino teeth, and others look like rodent teeth.

What is the hyrax's closest relative? A scientist would tell you that it is an elephant. Hyraxes don't look like elephants. But their front leg bones, feet, and brain are a lot like an elephant's.

Hyraxes and elephants are alike in other ways. Usually small mammals have their babies in about eight weeks. But the small hyrax takes over 7 months to have babies. Elephants also take a long time to have their babies.

Both hyraxes and elephants have trouble cooling down. An elephant will cool down in a river or pond. Or it will flap its ears to cool down.

Hyraxes keep cool by living on the boulders. If it gets too hot, they find a shady spot between the rocks. The hyraxes at Top-Knot can also go to a cave where the air never gets too warm.

Hyraxes have more to worry about than elephants do. The small hyraxes can make an easy meal for a larger animal. Since elephants are so big, they aren't easy to catch and eat.

Hyraxes are careful to look out for danger. One hyrax sits on the highest boulder. It's the guard. It watches the sky for hungry eagles and for hungry leopards in the grass. These animals like to eat hyraxes.

If the guard sees danger, it whistles. Then all the hyraxes dive into deep, dark spaces between the rocks. In seconds, the hyraxes disappear.

The island not only protects the hyrax, it also gives them food. In many parts of Africa, there is a dry season and a rainy season. In the dry season, with almost no rain, the grass turns brown and dies. But on Top-Knot, there are always green plants. In the morning, dew forms on the boulders and runs down into the dirt. This gives the plants the water they need to stay alive.

Many animals and birds come to Top-Knot to eat its green plants. One morning I was watching the hyraxes eat. Two giraffes came to the island looking for food. I was surprised to see the giraffes. The hyraxes were surprised too.

The giraffes stretched their long legs and necks. They stuck out their long tongues. But they couldn't reach the leaves they wanted. The green plants in Top-Knot were too low for them. The giraffes gave up and went to find a tree to nibble on.

After the giraffes left, the hyraxes nibbled on their morning food. A few minutes later, I heard some squeaks. The squeaks came from pygmy mongooses. These little animals wiggled, jumped, and rolled over. One stopped and I quickly took its picture.

Pygmy mongooses kill snakes. I felt safer knowing that. I had not seen a snake yet, and I hoped I would never see one.

The next day I saw three tiny hyraxes. They looked like tiny puppies, not like wild African animals. I didn't see their mother right away, but I knew she must be close by. Then I saw her on a rock. She was watching everything carefully.

The three tiny hyrax cubs got closer and closer to me. I could almost touch them. But then the mother decided they were getting too curious. She gave a soft call and the three cubs ran back to her side. I was glad she called them because people shouldn't touch wild animals. But they looked so soft that I wanted to touch them.

I spent the summer watching the hyraxes. I learned a lot about their habits. I learned why hyrax cubs are good at climbing rocks. They learn by climbing on their mothers. First they climb on her tail, and then they move up her back. They then reach her neck, and finally they sit on top of her head.

I saw one cub hop on his mother's back many times every day. He could jump up well but he wasn't so good at getting down. But he kept trying. Each time he jumped down, he got better at it. Then he became good at climbing up and down the rocks.

The hyraxes became used to having me in their territory. If I sat very still, they would come close. They nibbled flowers and leaves right near me. One day I saw something I hadn't noticed before. I saw that a hyrax had long hairs along its back. The hairs were stiff like the whiskers on a cat. Later, I found out that the hyraxes use them as "feelers." In caves and dark spaces, these feelers are like extra hands, helping them find their way.

Like most mammals, hyrax cubs sniff each other. I watched them do this many times. One cub sniffed me and then ran away. It looked back at me and seemed to ask, “Do you want to play?” Cubs often sniff adult hyraxes too. The adults nuzzle them and then send them off to play with the other cubs.

Adult hyraxes do not like to sniff each other face to face. If they do, they often end up fighting.

The adults spend most of their time away from each other or resting back to back. Resting like this helps the hyraxes in two ways. First, they won't fight because they aren't looking at each other. Second, they are looking in different directions and have a better chance of seeing danger.

In the afternoon the hyraxes rest in their cool dens. I took my lunch break while they rested. But one day, lunch had to wait. A very large rhinoceros was standing at the foot of Top-Knot. He was so close, I could see his ears. Rhinos can't see very well, so he didn't see me. I was lucky!

When the rhinoceros left, I went to my car to eat my lunch. Later, I thought I saw something move in a tree. When I looked more closely, a pair of yellow eyes stared back at me.

I wasn't the only one interested in the hyraxes. A long, lean leopard was stretched out on a big branch. He waited and watched. He was looking for a hyrax to snack on.

The hyraxes slowly came out of their dens. The leopard saw them and slowly climbed down the tree, moving quietly toward them.

Suddenly, one of the hyraxes whistled. At the same time, a herd of impalas, African antelopes, ran across the plains. The leopard turned, looked at them and ran after the impalas. An impala was a big meal, not just a snack!

It started getting dark. I walked back to my car and got in. Two hyrax cubs hopped onto the roof and peeked down at me. Then they jumped off. They ran back to the island to play tag on the rocks.

I drove away, thinking about hyraxes. They had escaped the leopard. I hoped they could escape most dangers. Hyraxes are small and easy to overlook, and that helps protect them. Their gray-brown fur helps them blend in with their background. They also have few enemies because people don't eat them or use their fur. Best of all, hyraxes have the perfect home.

My summer at the island ended. When I came to the islands, I wanted to learn about hyraxes. Now, everything on and around Top-Knot Island was special to me. The island was more than a pile of boulders. It was home to some wonderful animals. And of all the animals I had seen, the hyrax was my favorite.

When I left, the sun was going down. I took one last look at Top-Knot and said a silent goodbye.